Lara Morais

# Evaluation of stent surfaces after electropolishing using DOE

AF536378

Lara Morais

# Evaluation of stent surfaces after electropolishing using DOE

## Influence of process parameters on the surface appearance of coronary stents

ScienciaScripts

**Imprint**

Any brand names and product names mentioned in this book are subject to trademark, brand or patent protection and are trademarks or registered trademarks of their respective holders. The use of brand names, product names, common names, trade names, product descriptions etc. even without a particular marking in this work is in no way to be construed to mean that such names may be regarded as unrestricted in respect of trademark and brand protection legislation and could thus be used by anyone.

Cover image: www.ingimage.com

This book is a translation from the original published under ISBN 978-613-9-66902-8.

Publisher:
Sciencia Scripts
is a trademark of
Dodo Books Indian Ocean Ltd. and OmniScriptum S.R.L publishing group

120 High Road, East Finchley, London, N2 9ED, United Kingdom
Str. Armeneasca 28/1, office 1, Chisinau MD-2012, Republic of Moldova, Europe
Printed at: see last page
**ISBN: 978-620-8-11242-4**

Copyright © Lara Morais
Copyright © 2024 Dodo Books Indian Ocean Ltd. and OmniScriptum S.R.L publishing group

## SUMMARY

Cardiovascular diseases are treated by unblocking arteries and other peripheral pathways by applying a balloon accompanied by a stent to the obstructed area. Stents are small metal mesh cylinders that align themselves with the walls of blood vessels as soon as they are expanded by the balloon. The use of these coronary prostheses has been expanding in clinical use, since the surgery performed to treat such diseases is minimally invasive and guarantees less risk of restenosis. During the production of stents, a very important step takes place to give the surface of the stent a proper finish, which must be smooth and mirror-like to guarantee its stability and biocompatibility, called electropolishing. This process has several control parameters that influence the final surface and mass of the stent. Knowledge of these influences promotes a better understanding of each parameter in the surface quality and mass loss of the stent during electropolishing. One tool for properly analysing the ability of each parameter to influence a process is factorial planning, which makes it easier to obtain the composition of tests and then verify their results. This work aims to analyse the external surface of stents and their loss of mass after being electropolished by varying four parameters of this process, based on factorial design to set up the tests and verify the influence of each variable during reproduction. The results are intended to provide academic and industrial knowledge on unconventional surface finishing processes combined with statistical analysis of process variables.

Key words: Stent; Electropolishing; Factorial Planning; Co-Cr (L605).

## CONTENTS

# CHAPTER 1

## INTRODUCTION

This chapter presents a brief introduction to the general scenario of the object under study, its most relevant characteristics and some important considerations for researching and developing its applications.

### 1.1 Overview

Cardiovascular diseases are the leading cause of death in countries such as the United States and Brazil, with rates of 27 per cent and 32 per cent respectively. At the beginning of the 1990s, heart disease was only treated using a highly invasive method. This method consisted of performing surgery to implant a saphenous vein graft, i.e. replacing the blocked blood vessel with a healthy one. This procedure poses a high risk to the patient in the surgery itself and in the post-operative period, taking into account the high probability of rejection of the implanted vessel (ABELIN, 2009; ARAÚJO, 2007).

More than a century ago, an English dentist called Charles Stent devised a dental material for moulding, which was later used as a support for living tissue in healing. In 1964, Dotter and Judkins described the first angioplasty procedure using a dilatation catheter in peripheral circulation, foreshadowing its use in coronary circulation. In 1977, in Switzerland, Andréas Gruentzig, a young German doctor working at a university hospital in Zurich, developed a dilatation catheter with a double lumen and a balloon made of a non-elastic material, which he used successfully in femoral arteries. Later that year, Gruentzig performed the first Percutaneous Transluminal Coronary Angioplasty (PTCA) on a man, expanding its use to coronary arteries, which gave a great boost to invasive cardiology and interventionism (ARAÚJO, 2007).

The implantation of *stents* for the treatment of coronary obstruction has increased the safety of this surgical intervention, proving to be efficient in resolving the problem in the medium term. The greater acceptance of the use of *stents* as a primary modality for the percutaneous treatment of coronary atherosclerotic disease has expanded their applications to more complex lesions. Combined with technological growth, this prompted an immediate response from the industry, leading to the development of a wide variety of vascular *stent* models, with innovative designs and characteristics, becoming the dominant modality for percutaneous coronary intervention worldwide (CHAMIÉ, 2009; COSTELLA, 2010).

The majority of coronary *stents* tested clinically are made from metal alloys, with stainless steel being the most frequently used. Steel is predominantly composed of iron, which is biologically inert, but it also contains around 12% nickel and 2% molybdenum, substances which, in allergic individuals, can contribute to an increased risk of *in-stent* restenosis. Cobalt chromium alloys have made it possible to manufacture *stents* with thinner struts without jeopardising their radiopacity and radial strength. Characteristics of the *stent* surface can also influence the performance of these prostheses (CHAMIÉ, 2009).

In Brazil, by mid-2009, around 108 *stents* had been registered with the National Health Surveillance Agency (ANVISA) for clinical use, none of which met the criteria for an "ideal *stent*" but many of which are constantly evolving. Among the desirable characteristics for *stents* are high longitudinal flexibility, radial elasticity, radiopacity, minimal wrinkling after implantation, easy retrieval in the event of implantation failure, among others. In addition, the existence of a low thrombogenic surface, which slows the growth of arterial tissue around the implant, is fundamental for the proper performance of these endoprostheses. In this sense, a good surface treatment, resulting in low roughness, is essential for reducing the risks inherent in this intervention. The biostability of *stents* is achieved by appropriate surface finishing processes, such as electropolishing and acid passivation (COSTELLA, 2010).

The final finish of a metal surface subjected to an electropolishing process depends on the prevailing current distribution conditions and, therefore, on geometric, electrical and hydrodynamic parameters (VEROLI, 2011). The parameters used in the electropolishing process must be optimised in order to obtain the best possible results.

Optimising the process depends on good test planning to investigate it. For such an investigation, the general procedure is to formulate hypotheses and verify them directly or by their consequences. This requires a set of observations and the design of experiments is then essential to indicate the scheme under which the hypotheses can be verified. The hypotheses are verified using statistical analysis methods that depend on the way in which the observations were obtained. Therefore, planning experiments and analysing the results are closely linked and should be used in sequence in scientific research in the various areas of knowledge (MARINHO, 2005).

Given the importance of *stents,* it is necessary to investigate the processes involved in their manufacture, such as electropolishing. Its study depends on good test planning so that the process parameters can be analysed and the hypotheses under their influence understood. The results obtained in the tests carried out using factorial planning serve as a basis for a better understanding of the electropolishing process for application to Co-Cr *stents* and also for further studies on the subject.

## 1.2 - Defining the problem and objectives

In industrial manufacturing processes there are various factors and levels of regulation that influence the quality characteristics of products and a common problem encountered by companies when carrying out experiments is the need to simultaneously study the effect of these factors with different levels of regulation.

The process of electropolishing *stents,* for example, depends on multiple variables that directly or indirectly influence the final surface finish. The adjustment of these variables (parameters) must be defined in such a way as to understand the effect of each one during the process, bearing in mind that the geometric arrangements of the equipment also influence the result. The best combination of parameters must be the one that adapts the material *(stent)* to the specifications of surface integrity and dimensional tolerance.

In this sense, the general objective of this work is:

- To visually and dimensionally evaluate the external surface of *stents* and their consequent loss of mass after the electropolishing process, using factorial planning as a tool to determine the influence of the test parameters.

The specific objectives are:

- Cut, strip and heat-treat the Co-Cr *stents*;
- Measure the thickness of the *stents* and check their weight;
- Set up the factorial design to carry out the tests;
- Adjust the electro-polishing equipment (current, time, number of cycles and speed);
- Carry out tests with the different combinations of parameters obtained from the factorial design;
- Carry out a visual analysis of the external surface of the *stents,*
- Measure the thickness of the *stents,*
- Check the weight of the electro-polished *stents*;
- Evaluate the influence of the parameters on the visual, dimensional and mass loss analyses.

# CHAPTER 2

## LITERATURE REVIEW

This chapter presents a review of the scientific papers studied with regard to the origin of biomaterials, the manufacture of *stents* and their surface finishing process, also known as electropolishing, the use of design of experiments and the analysis of their results. All this information contributes to the study and development of the proposed work.

### 2.1 - **Biomaterials**

The most widely accepted definition of biomaterials today is that of the Consensus Conference (1982), which defines a biomaterial as any substance or combination thereof, with the exception of drugs or pharmaceuticals, of natural or synthetic origin, which can be used for any length of time, partially or totally augmenting or replacing any tissue, organ or function of the organism (COSTELLA, 2010).

According to Costella (2010), the clinical acceptance of an implant material must fulfil some fundamental requirements:

- The material must be biocompatible, i.e. its presence must not cause harmful effects at the implant site or in the biological system;
- Fabrics should not cause material degradation, such as corrosion on metals, or at least to a tolerable extent;
- The material must be biofunctional, i.e. it must have the right mechanical characteristics to fulfil the desired function for the desired length of time;
- The material must be sterilisable.

The main advances in the field of biomaterials are due to the increase in population and life expectancy. Currently, a biomaterial is a substance that has been designed to have a form that, independently or as part of a complex system, is used to direct, by controlling interactions with components of the living system, the progress of any therapeutic or diagnostic procedure in human or veterinary medicine.

## 2.2 - **Metallic biomaterials**

Steel has been used since the 19th century as plates and screws for fracture fixation. Fixation with screws provided a stronger fixation than the previous method, with metal wires. Problems related to corrosion and toxicity led to the development of new alloys with greater resistance to corrosion. Consequently, stainless steel, Co-Cr alloys, titanium and titanium alloys gradually became the most widely used materials in implants (COSTELLA, 2010).

The biostability of the material is achieved by appropriate surface finishing processes, such as electropolishing and acid passivation. These surface treatments are widely used to finalise the *stent* manufacturing process and significantly improve the stability of the materials, reducing the risks associated with implantation. The electropolishing process removes the original naturally formed oxide layer, replacing it with a smoother, more homogeneous layer with nanometre-scale roughness (AIHARA, 2009).

## 2.3 - **Stainless steel**

The stainless steel most commonly used as a biomaterial is 316L stainless steel, which is austenitic and hardens by hardening. This type of stainless steel has a low carbon content (0.03% max.), which gives it greater resistance to corrosion in saline and chloride-rich environments (characteristics of the physiological environment). The nickel in its composition is responsible for austenite stabilisation and corrosion resistance. Its high resistance to corrosion is due to its high chromium content. Levels of this element above 28% allow the precipitation of Cr $3C_{26}$ on grain boundaries, which would be prefrerential sites for intergranular corrosion (COSTELLA, 2010).

The main limitations of using stainless steel implants are corrosion and ion exchange. Stainless steels undergo corrosion in vivo and release ions such as $Ni^{2+}$ , $Cr^{3}$

* and $Cr^6$ *, which can cause local (irritation, inflammation in areas adjacent to the implant-tissue interface) and systemic effects (COSTELLA, 2010).

## 2.4 -Co-Cr alloy (L605)

Cobalt-based alloys were first investigated by Elwood Haynes in the mid-1900s. During the First World War, tools made from cobalt were commonly used and in 1922 cobalt alloys with high wear resistance were used for ploughs, oil well drill bits and in internal combustion engines. In the early 1940s cobalt-based alloys with high temperature and corrosion resistance were introduced, so many of the applications for the L605 alloy were for gas turbines in power stations. Other cobalt-based alloys have a long history in orthopaedic, cardiac and dental applications (AIHARA, 2009).

The L605 alloy has high ductility, corrosion resistance and greater structural stability than wear-resistant forged alloys. The alloy's high elastic modulus and high density are factors that encourage its application in *stents*. The presence of precipitates in this alloy, due to the physical metallurgy of the system, makes it difficult to predict the behaviour of the implanted material. The real influence of these precipitates is still unknown at low temperatures and they are considered undesirable. The precipitates present in L605 have a direct effect on the surface finish and can affect coating adhesion, the balloon/stent interface, particle release into the bloodstream, etc. (AIHARA, 2009).

## 2.5 -Coronary Angioplasty and *Stents*

Angioplasty can be defined simply as the process of unblocking fatty plaques that have accumulated inside the arteries (stenosis) by inserting a flexible tube called a catheter with a small balloon at the end. This procedure is carried out through a section in the groin area. The catheter is guided to the area of the artery previously blocked by fat. Figure 2.1 shows the aforementioned definition in simplified form (ARAÚJO,

2007).

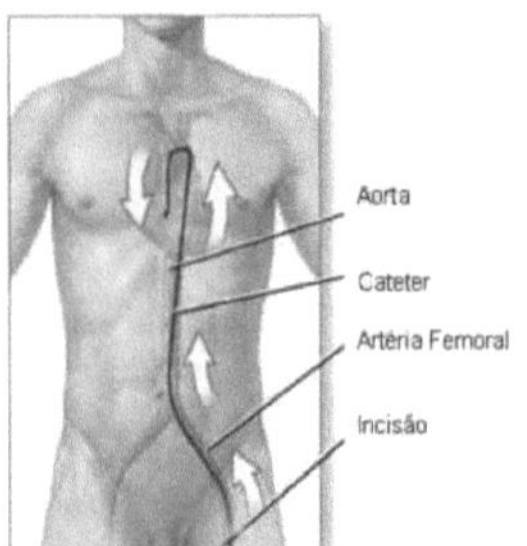


FIGURA 2.1 - Coronary angioplasty procedure illustrated in simplified form (modified from ARAÚJO, 2007).

*Stents* consist of mesh-like tubular metal structures inserted into the vascular channel to keep it open by means of mechanical pressure, normalising blood and oxygen flow in the channel (COSTELLA, 2010).

There are two types of *stents* on the market today: self-expanding and balloon-expandable. Self-expanding stents, as their name suggests, do not require a balloon to expand. These *stents* are generally made from a Ni-Ti metal alloy, also known as Nitinol. This shape memory alloy has excellent recovery after deformation, up to 8% more than other metal alloys, such as Cu-based alloys (copper - 4% to 5%), and also offers a high level of ductility and excellent resistance to corrosion. Figure 2.2 illustrates a self-expanding *stent* and how it exits the catheter (ARAÚJO, 2007).

(a) (b)

FIGURE 2.2 - NiTi *stent* (a) and the way the *stent* leaves the catheter (b) expanding as it is released (NORMAN NOBLE, 2013; MEDINTPRO, 2014).

Balloon-expandable *stents* are commonly made from 316L stainless steel, but recently studies have indicated that this material has been replaced by a metal alloy based on cobalt-chromium. This new alloy has greater resistance and better biocompatibility (ARAÚJO, 2007). These *stents* are placed with greater precision and have good resistance to radial compression, but have little elasticity and longitudinal flexibility. They are indicated for sites prone to calcification and external compression, and are contraindicated for sites close to joints, around bones and ligaments and near flexion points. These *stents* normally open from the tips towards the centre, so that they attach to the wall of the blood vessel. Figure 2.3 illustrates a balloon-expandable *stent* and how it is positioned in the artery (COSTELLA, 2010).

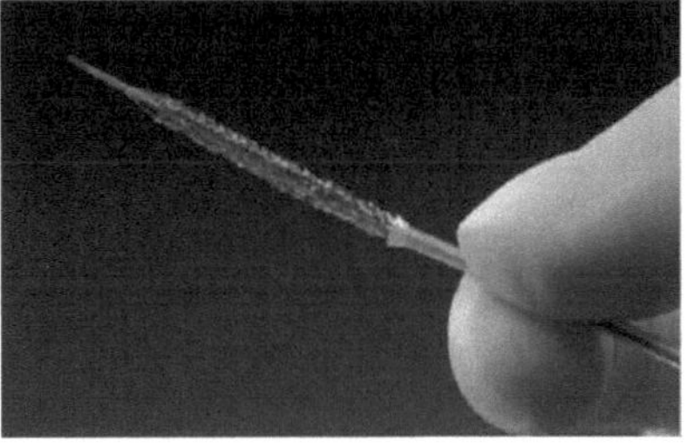

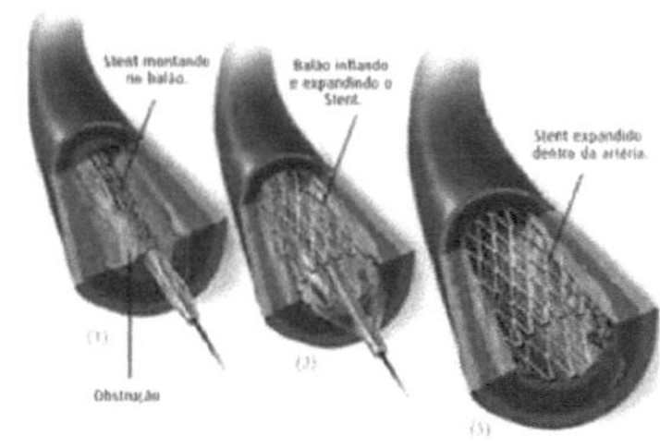


(a) (b)

FIGURE 2.3 - Balloon-expandable *stent* (a) and positioning inside the fat-blocked artery (b) (COSTELLA, 2010; OLIVEIRA, 2010).

## 2.6 - Electropolishing

According to the ASTM B374-06 standard (2011), the electropolishing process is defined as improving the surface finish of a metal by imposing a positive (anodic) potential/current in an appropriate solution. Verolli (2011) states that the electropolishing process can be used for the following purposes:

- Removal of deformations on the metal surface, such as non-metallic inclusions, is an important tool in the manufacture of nuclear equipment, high pressure vessels and jet engine vanes;
- Electropolishing of reactors, pipes, installations, valves, heat exchangers, making these materials more resistant to corrosion and with a flatter, easily cleanable surface;
- Removal of burrs from metal machines after cutting or milling processes,

thinning of tools, checking the dimensions of measuring instruments and flattening of parts subject to friction.

In the electropolishing process, usually carried out in acidic solutions, anodic currents or potentials induce the dissolution or passivation of the metal, promoting the levelling, gloss and specular reflectivity of the metal surface due to the decrease in its roughness at a micrometric level. The electrochemical characteristics of electropolishing are such that the metal cannot passivate, but neither can it undergo accelerated dissolution. The final finish of a metal surface subjected to an anodic dissolution process depends on the prevailing current conditions and therefore on geometric, electrical and hydrodynamic parameters (VEROLI, 2011).

## 2.7 - **Factorial Planning**

The designed or planned experiment is a test or series of tests in which deliberate changes or stimuli are induced in the *input* variables of the process or system, in such a way that it is possible to observe and identify the effects on the responses or output variables. The transformation process or system is represented by the combination of machines, methods, people and other resources that transform an input into finished or semi-finished products with specific characteristics or parameters (GALDAMEZ, 2002). The simplification of the transformation system is exemplified in Figure 2.4.

FIGURE 2.4 - General model of a transformation system (GALDAMEZ, 2002).

The application of experimental planning in industry is fundamental to the development of new products and the control of processes. In this area it is common to encounter problems in which several properties need to be studied at the same time and these, in turn, are affected by a large number of experimental factors. The role of design of experiments techniques is to help manufacture products with better characteristics, reduce development time, increase process productivity and minimise sensitivity to external factors (POMPEU, 2013).

In order to better understand the fundamentals of factorial planning, it is necessary to have a good grasp of some basic knowledge (CUNICO et al, 2008), described below:

-Factor : each variable of the system under study.

-Level : operating conditions of the control factors investigated in the experiments. They are generally identified by low level (-) and high level (+).

-Statistical model: model of the type y=b0+blxl where bO, bl ...., bn are the effects of each factor on the response.

-Effect : change that occurs in the response when changing from the low level (-) to the high level (+)-

To carry out a factorial design, we first need to specify the levels at which each factor will be studied, i.e. the factor values (or versions, in qualitative cases) that will be used in the experiments. Each of these experiments, in which the system is subjected to a defined set of levels, is an experimental test. In general, if there are nj levels of factor 1, $n_2$ of factor 2,..., and $n_k$ of factor k, the design will be a factorial x $n_2$ x ... x $n_k$ . This does not necessarily mean that only $n_1$ x *n? x... experiments* will be carried out. *x* $n_k$ experiments. This is the minimum number to have a complete factorial design. The experimenter may want to repeat tests to get an estimate of the experimental error, in which case the number of experiments will be greater (MARINHO, 2005).

## 2.8 - Factorial Planning $2^k$

Two-level full factorial design or 2-factor design$^k$ is the type of design in which two levels of values are set for each factor, high and low, and each combination of factors is tested. One aspect to consider in this type of design is that, as there are only two levels of each factor, it has to be assumed that the response is approximately linear over the range of factor levels chosen. Another important aspect is that for experiments with a large number of factors being considered, the full factorial results in an extremely large number of combinations to be tested. In this situation, fractional design is used to strategically select a subset of combinations to test in order to identify the factors with little or no importance in the system's performance (FILHO, 2006).

To illustrate the procedure of this technique, consider an experiment with three factors (xl, x2 and x3), each of these parameters was tested with two levels (-1, +1). The planning matrix for the 2-factor experiment[3] is shown in Table 2.1. It is important to note that the order in which the test is carried out is defined randomly (GALDAMEZ, 2002).

Table 2.1 - Test matrix for a 2 plan[3] (GALDAMEZ, 2002).

| Ne Test | Control factors | | | Test order | Answer (yi) |
|---|---|---|---|---|---|
| | xl | x2 | x3 | | |
| 1 | -1 | -1 | -1 | 6 | yi |
| 2 | +1 | -1 | -1 | 8 | y2 |
| 3 | -1 | +1 | -1 | 1 | y3 |
| 4 | +1 | +1 | -1 | 2 | y4 |
| 5 | -1 | -1 | +1 | 5 | ys |
| 6 | +1 | -1 | +1 | 3 | y6 |
| 7 | -1 | +1 | +1 | 4 | y7 |
| 8 | +1 | +1 | +1 | 7 | y8 |

## 2.9 - Analysing results

*Statistical software (Statistica, Scilab, ExceT)* can be used at this stage to help use the techniques for planning and analysing experiments, linear graphs and normal probability graphs. Statistical concepts are applied to the results of an experiment to describe the behaviour of the control variables, the relationship between them and to estimate the effects produced on the observed responses (GALDAMEZ, 2002).

The effects of each variable on the process are then analysed. The effects can be classified into two categories: main effect (effect relating to the change in level of a single factor) and interaction effect (effect relating to the change in level between two or more factors at the same time). In order to calculate the effects, in addition to the coding for the individual effects, the same is required for the interaction effects. The sign of the effect of an interaction between factors is obtained by multiplying the signs of the factors involved in the interaction (MOREIRA, 2009). Table 2.2 shows the interaction of three factors.

Table 2.2 - Interaction between three factors in an experiment $2^3$ (MOREIRA, 2009).

| **Experiments** | **Effects** | | | | | | |
|---|---|---|---|---|---|---|---|
| | **Control factors** | | | **Interactions** | | | |
| **Test No.** | xl | x2 | x3 | **xl-x2** | **xl-x3** | **x2-x3** | **xl-x2-x3** |
| 1 | -1 | -1 | -1 | +1 | +1 | +1 | -1 |
| 2 | +1 | -1 | -1 | -1 | -1 | +1 | +1 |
| 3 | -1 | +1 | -1 | -1 | +1 | -1 | +1 |
| 4 | +1 | +1 | -1 | +1 | -1 | -1 | -1 |
| 5 | -1 | -1 | +1 | +1 | -1 | -1 | +1 |
| 6 | +1 | -1 | +1 | -1 | +1 | -1 | -1 |
| 7 | -1 | +1 | +1 | -1 | -1 | +1 | -1 |
| 8 | +1 | +1 | +1 | +1 | +1 | +1 | +1 |

After analysing the data, the effects and their significance, a detailed statistical analysis is then carried out to obtain the respective deviations and variances during the tests. These calculations allow the researcher to understand the variation in the system's output when some input variable (whether directly or indirectly significant) is changed. This work comprises a 2 factor design$^{4}$ , i.e. four variables were studied at two levels (maximum and minimum) to understand their influence on the external surface finish and mass loss after electropolishing 19mm long *stents*.

# CHAPTER 3

## METHODOLOGY

The methodology consisted of carrying out electropolishing tests on Co-Cr balloon expandable *stents,* using a tool for statistical planning and data processing *(Scilab),* laser cutting, chemical stripping, heat treatment, electropolishing, a profile projector to measure the thickness of the *stents* (before and after they were electropolished), an optical microscope to visually analyse the *stents* and a precision scale to measure the weight of the *stents* (before and after they were electropolished). The tests were carried out in a laboratory located in a medical products industry, which for reasons of integrity will not be named.

The equipment manuals were read and a bibliographical review of the subject was carried out based on the characteristics of the electropolishing process and the ways of carrying out a 2 factorial design$^k$ . The tests were carried out in compliance with occupational health and safety regulations.

### 3.1 - **Preliminary tests**

Preliminary electropolishing tests were carried out in order to obtain the significant process variables, as well as their maximum and minimum levels. Figure 3.1 shows the machine used for the electropolishing process. The range of parameter values was chosen so that the minimum and maximum levels guaranteed that polishing would occur regardless of whether or not the result was within acceptable tolerance standards.

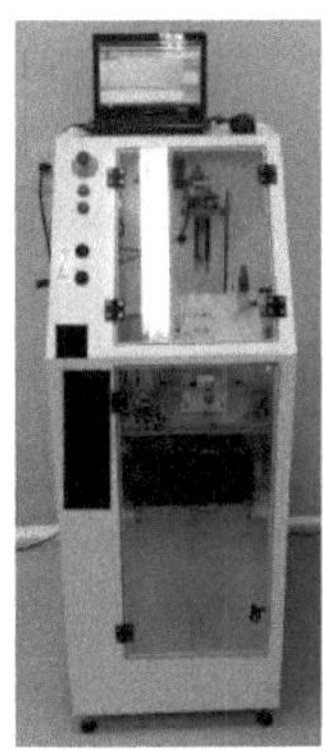

FIGURA 3.1 - Electropolishing machine.

Based on these tests, a 2 factor design was carried out[4] , where the factors used were: current, cycle time, number of cycles and translation speed. Figure 3.2 illustrates the transformation system for this design and Table 3.1 shows the coding of the design matrix.

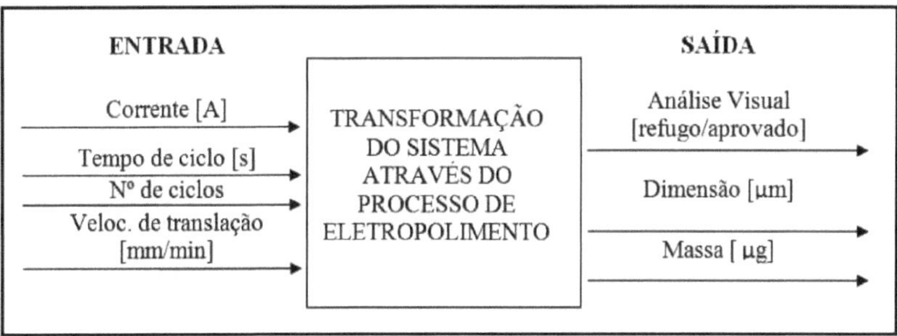


FIGURA 3.2 - Representation of the transformation system realised in this work.

The transformation system in question (Figure 3.2) is the electropolishing process. The input variables are the parameters that were modified throughout the tests following the factorial design. The output, i.e. the response through which the influence of each parameter was analysed, is the visual analysis of the *stent,* the dimensional analysis (reduction in thickness) and the loss of mass.

Table 3.1 - Coding for the planning matrix.

| Level | Factors | | | | | | | |
|---|---|---|---|---|---|---|---|---|
| | Current [A] | | Cycle time [s] | | Number of cycles | | Travel speed [mnimin] | |
| Maximum | 1 | 2 | 1 | 60 | 1 | 10 | 1 | 100 |
| Minimum | -1 | 0 6. | -1 | 10 | -1 | 2 | -1 | 0 |

In order to maintain the integrity and industrial confidentiality of the company in which the work was carried out, the values of the parameters used in the tests will not be specified, but rather the range in which they were carried out. The machine's electric current operates in a range from 0.1 A to 10A, the cycle time can be chosen from ls to 3600s, the number of cycles has no specific limit and the speed of translation of the electrode supporting the *stent* (Figure 3.3) in the acid solution varies from Omm/min to 800mm/min. In view of this and the preliminary tests, the range of values to be used for the development of the work can be chosen: electric current ranging from 0.6A to 2A, cycle time ranging from lOs to 60s, translation speed from Omm/min to lOOmm/min and cycle number between 2 and 10 cycles. It is worth emphasising again that these values are not necessarily the ones used for the tests, as this is a very specific detail within a production process developed within the company.

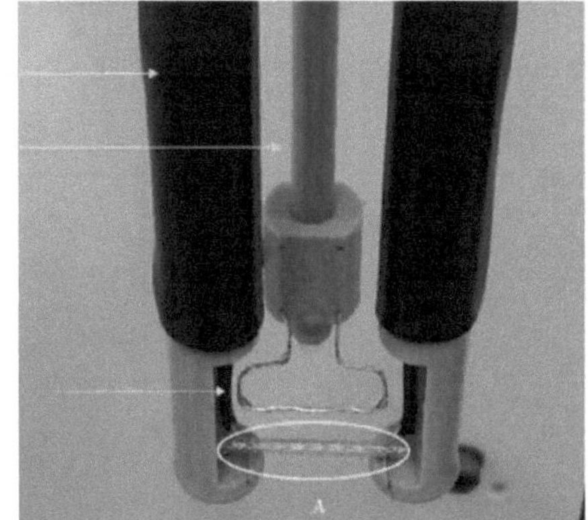


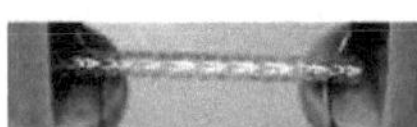


FIGURA 3.3 - Electrode tower (a) supporting the *stent* (b).

The electrode tower (Figure 3.3) is the positive pole of the electropolishing process and holds the *stent* when it is immersed in the acid solution contained in the machine (see Figure 3.1). In this container there are two metal plates situated at such a distance that the tower is exactly in the middle between them during electropolishing. The plates are the negative pole of the process and stabilise the electric field generated during the passage of electric current through the *stent* in the acid solution. The tower can move in a vertical direction during the process and its translation speed can be

0mm/min (stationary) or any other value up to 800mm/min, as mentioned above.

## 3.2 - **Final tests**

The final tests were carried out on the basis of the working ranges of the most significant parameters found in the preliminary tests. The planning matrix (Table 3.2) was developed from the coding of the matrix in Table 3.1.

Table 3.2 - Planning matrix.

| Essay | Main effects | | | |
|---|---|---|---|---|
| | C | TC | NC | VT |
| 1 | -1 | -1 | -1 | -1 |
| 2 | 1 | -1 | -1 | -1 |
| 3 | -1 | 1 | -1 | -1 |
| 4 | 1 | 1 | -1 | -1 |
| 5 | -1 | -1 | 1 | -1 |
| 6 | 1 | -1 | 1 | -1 |
| 7 | -1 | 1 | 1 | -1 |
| 8 | 1 | 1 | 1 | -1 |
| 9 | -1 | -1 | -1 | 1 |
| 10 | 1 | -1 | -1 | 1 |
| 11 | -1 | 1 | -1 | 1 |
| 12 | 1 | 1 | -1 | 1 |
| 13 | -1 | -1 | 1 | 1 |
| 14 | 1 | -1 | 1 | 1 |
| 15 | -1 | 1 | 1 | 1 |
| 16 | 1 | 1 | 1 | 1 |

*C - current, CT - cycle time, NC - number of cycles and VT - travel speed.

Table 3.2 shows the maximum (1) and minimum (-1) levels for each of the 16 tests established by the 2 factorial design[4,] for each parameter. From this table, the other calculations intrinsic to the factorial design can be developed. The effects relating to the responses (outputs) obtained after the tests and their replicates were established. Each test was carried out 3 (three) times, totalling 48 (forty-eight) tests, which were randomised.

## 3.3 - **Data processing in *Scilab***

Based on the tests and replicates carried out, a *script* (Appendix A) was drawn up for the calculations inherent in the full factorial design ($2^4$ ) to create a mathematical model capable of predicting dimensional behaviour (thickness), mass loss and polishing (weighted visual analysis), considering the range of variables proposed in this work. The significance of each parameter involved was also analysed, along with the interaction between them.

# CHAPTER 4

## RESULTS

This chapter presents the results obtained from the electropolishing of 48 (forty-eight) *stents* using factorial planning as a tool to set up the experiments and *Scilab software* to analyse the data generated at the end of the process.

### 4.1 - Structuring the contrast matrix for planning $2^4$

From the preliminary tests, the most relevant process parameters can be seen, as well as the maximum and minimum levels for each parameter. Table 4.1 shows the structure of the contrast matrix for planning $2^4$ .

Table 4.1 - Contrast matrix with main effects and interactions between parameters.

| Essay | Main effects | | | | Int. 2° order | | | | | | Int. 3rd order | | | | Int. 4° order |
|---|---|---|---|---|---|---|---|---|---|---|---|---|---|---|---|
| | C | TC | NC | VT | C TC | C NC | C VT | TC NC | TC VT | NC VT | C TC NC | C TC VT | C NC VT | TC NC VT | C TC NC VT |
| 1 | -1 | -1 | -1 | -1 | 1 | 1 | 1 | 1 | 1 | 1 | -1 | -1 | -1 | -1 | 1 |
| 2 | 1 | -1 | -1 | -1 | -1 | -1 | -1 | 1 | 1 | 1 | 1 | 1 | 1 | -1 | -1 |
| 3 | -1 | 1 | -1 | -1 | -1 | 1 | 1 | -1 | -1 | 1 | 1 | 1 | -1 | 1 | -1 |
| 4 | 1 | 1 | -1 | -1 | 1 | -1 | -1 | -1 | -1 | 1 | -1 | -1 | 1 | 1 | 1 |
| 5 | -1 | -1 | 1 | -1 | 1 | -1 | 1 | -1 | 1 | -1 | 1 | -1 | 1 | 1 | -1 |
| 6 | 1 | -1 | 1 | -1 | -1 | 1 | -1 | -1 | 1 | -1 | -1 | 1 | -1 | 1 | 1 |
| 7 | -1 | 1 | 1 | -1 | -1 | -1 | 1 | 1 | -1 | -1 | -1 | 1 | 1 | -1 | 1 |
| 8 | 1 | 1 | 1 | -1 | 1 | 1 | -1 | 1 | -1 | -1 | 1 | -1 | -1 | -1 | -1 |
| 9 | -1 | -1 | -1 | 1 | 1 | 1 | -1 | 1 | -1 | -1 | -1 | 1 | 1 | 1 | -1 |
| 10 | 1 | -1 | -1 | 1 | -1 | -1 | 1 | 1 | -1 | -1 | 1 | -1 | -1 | 1 | 1 |
| 11 | -1 | 1 | -1 | 1 | -1 | 1 | -1 | -1 | 1 | -1 | 1 | -1 | 1 | -1 | 1 |
| 12 | 1 | 1 | -1 | 1 | 1 | -1 | 1 | -1 | 1 | -1 | -1 | 1 | -1 | -1 | -1 |
| 13 | -1 | -1 | 1 | 1 | 1 | -1 | -1 | -1 | -1 | 1 | 1 | 1 | -1 | -1 | 1 |
| 14 | 1 | -1 | 1 | 1 | -1 | 1 | 1 | -1 | -1 | 1 | -1 | -1 | 1 | -1 | -1 |
| 15 | -1 | 1 | 1 | 1 | -1 | -1 | -1 | 1 | 1 | 1 | -1 | -1 | -1 | 1 | -1 |
| 16 | 1 | 1 | 1 | 1 | 1 | 1 | 1 | 1 | 1 | 1 | 1 | 1 | 1 | 1 | 1 |

* C - current, TC - cycle time, NC - number of cycles and VT - travel speed.

Table 4.1 shows the maximum (1) and minimum (-1) levels of the parameters used for the electropolishing process, namely current (C), cycle time (TC), number of cycles (NC) and translation speed (VT). The parameter interactions are:

a) Second-order, i.e. the interaction of two parameters:

- Current and cycle time (CTC);
- Current and number of cycles (CNC);
- Chain and travel speed (CVT);
- Cycle time and number of cycles (TCNC);
- Cycle time and travel speed (TCVT);
- Number of cycles and speed of travel (NCVT);

b) Third-order, i.e. the interaction of three parameters:

- Current, cycle time and number of cycles (CTCNC);
- Current, cycle time and travel speed (CTCVT);
- Current, number of cycles and travel speed (CNCVT);
- Cycle time, number of cycles and travel speed (TCNCVT);

c) Fourth-order, i.e. interaction between all four parameters:

- Current, cycle time, number of cycles and travel speed (CTCNCVT).

## 4.2 - Preparing the specimens and carrying out the tests

The 1.8mm diameter Co-Cr tube was cut to produce the test *stents* (Figure 4.1). The *stents* were then washed in acid (pickling) to remove all oxide and burrs from the cut. The *stents* were then heat treated to relieve tension.

**FIGURE 4.1 - Laser cutting of *stents* (ENGENHARIA BIOMÉDICA, 2010).**

## 4.3 - Preliminary measurements

**After undergoing all the processes mentioned in item 4.2, the *stents* had their thicknesses measured using a profile projector (Figure 4.2), and their weights measured on a high-precision scale. This data served as a basis for comparing the results after electropolishing.**

FIGURE 4.2 - *Stent* in the profile projector.

## 4.4 - Electropolishing tests carried out

The electropolishing tests were carried out following the schematisation of the maximum and minimum levels according to the planning matrix (Table 3.2). The order in which the tests were carried out was randomised and is illustrated in Table 4.2.

Table 4.2 - Order of tests.

| Trial order (planning) | Running order (random) |
|---|---|
| 1 | 10 |
| 2 | 14 |
| 3 | 8 |
| 4 | 16 |
| 5 | 7 |
| 6 | 3 |
| 7 | 5 |
| 8 | 13 |
| 9 | 11 |
| 10 | 4 |
| 11 | 6 |
| 12 | 12 |
| 13 | 15 |
| 14 | 1 |
| 15 | 9 |
| 16 | 2 |

Randomising the tests as illustrated in Table 4.2 allows the effects of uncontrolled factors, which affect the response variable and may be present during the experiment, to be balanced between all the possible measurements. This balance avoids possible confounding in the evaluation of results due to the action of these factors.

## 4.5 - Final measurements and visual analysis

After undergoing the electropolishing process, the thicknesses of the *stents* were measured again using a profile projector (Figure 4.2), their weights were measured on a high-precision scale located in a clean room with controlled pressure and their external surface was analysed according to the level of polish. This level of polishing was weighted (Table 4.3) so that the number 1 represents a poor level of polishing, i.e. with little or no change in gloss (Figure 4.3), the number 3 represents an intermediate level of polishing in which the surface is shiny but not completely polished (Figure 4.4) and the number 5 represents an ideal level of gloss with adequate polishing (Figure 4.5).

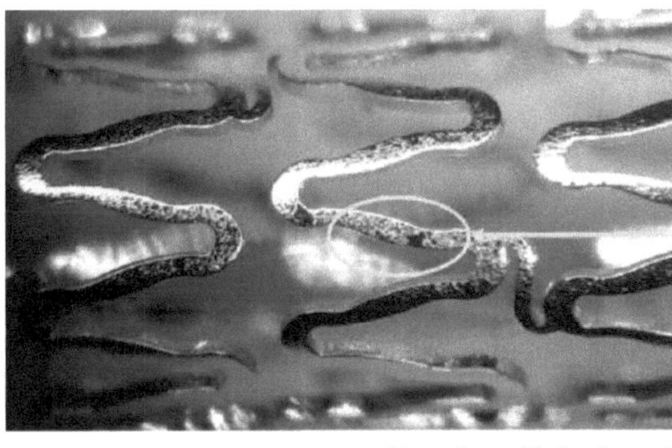

Superfície com arranhões e contornos (rugosidade).

Figura 4.3 - **Poorly polished surface (weighting 1).**

**Figure 4.3 illustrates the external surface of a *stent* with poor polishing. You can see the scratches and contours that appear as surface roughness on the *stent*. The *stents* analysed under the microscope that showed the same level of surface integrity as the *stent* in Figure 4.3 were considered to be poorly polished and given the number 1 for the visual analysis weighting.**

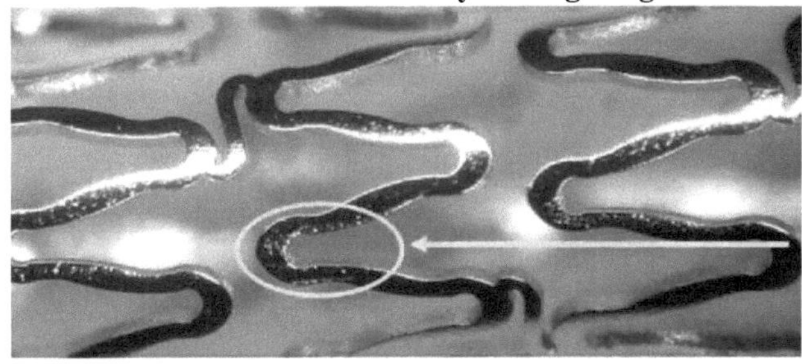

Superfície polida, mas com riscos pouco profundos.

Figura 4.4 - **Surface with intermediate polish (weighting 3).**

**Figure 4.4 illustrates the external surface of a *stent* with intermediate polishing. It is possible to see some shallow scratches that appear as slight surface roughness on the *stent*. The *stents* analysed under the microscope that showed the**

**same level of surface integrity as the *stent* in Figure 4.4 were considered to have intermediate polishing and were given the number 3 for the visual analysis weighting.**

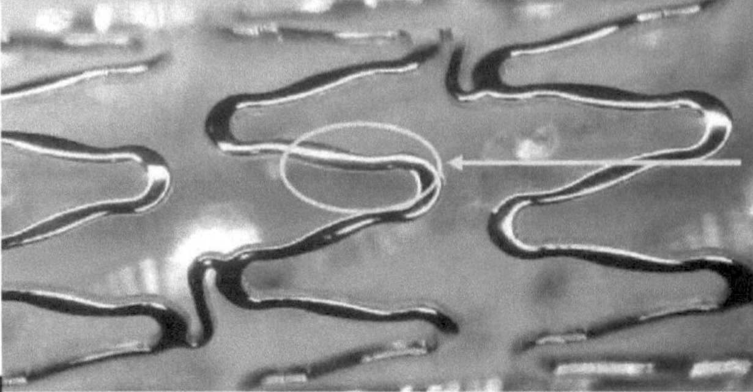


Figura 4.5 - **Properly polished surface (weighting 5).**

**Figure 4.5 illustrates the external surface of a well-polished *stent*. You can see that scratches, scuffs or contours are no longer present on the surface of the *stent*. The *stents* analysed under the microscope that showed the same level of surface integrity as the *stent* in Figure 4.5 were considered to be adequately polished and given the number 5 for the visual analysis weighting.**

**Table 4.3 - Results of the visual analysis.**

| Essay | Weighting the visual analysis | | |
|---|---|---|---|
| 1 | 3 | 1 | 1 |
| 2 | 3 | 3 | 1 |
| 3 | 1 | 3 | 1 |
| 4 | 3 | 3 | 3 |
| 5 | 1 | 1 | 1 |
| 6 | 3 | 3 | 3 |
| 7 | 5 | 5 | 5 |
| 8 | 5 | 5 | 5 |
| 9 | 1 | 1 | 3 |
| 10 | 3 | 3 | 1 |
| 11 | 3 | 1 | 1 |
| 12 | 3 | 5 | 5 |
| 13 | 1 | 1 | 1 |
| 14 | 3 | 3 | 3 |
| 15 | 5 | 5 | 5 |
| 16 | 5 | 5 | 5 |

Table 4.3 shows the results of the visual analysis according to the weighting described above. The test column refers to the order of the factorial design (Table 3.2) and each test was carried out three times (replicates), so there were three results to observe.The results obtained from the dimensional analysis (decrease in thickness) and loss of mass of the 48 electropolished *stents* are shown in Table 4.4. As the *stent* is a very small object (19mm long) its dimensions are in pm and its mass is in μg.

Table 4.4 - Thickness reduction and mass loss results for electropolished *stents*.

| Essay | Thickness decrease [μm] | | | Mass loss [μg] | | |
|---|---|---|---|---|---|---|
| 1 | 3,7 | 2,7 | 2,0 | 2174 | 2281 | 2308 |
| 2 | 2,0 | 3,3 | 2,0 | 3173 | 3260 | 3347 |
| 3 | 4,3 | 3,3 | 8,0 | 4674 | 4290 | 5194 |
| 4 | 10,0 | 11,0 | 6,3 | 7051 | 7503 | 6940 |
| 5 | 6,7 | 5,3 | 5,3 | 5220 | 3825 | 4516 |
| 6 | 8,7 | 9,0 | 5,7 | 6633 | 6029 | 6922 |
| 7 | 10,0 | 7,3 | 12,7 | 9127 | 9153 | 9217 |
| 8 | 18,0 | 16,7 | 13,7 | 14898 | 13921 | 14007 |
| 9 | 3,7 | 4,0 | 2,3 | 2232 | 2337 | 2400 |
| 10 | 6,0 | 3,7 | 10,0 | 3688 | 3248 | 3060 |
| 11 | 6,0 | 4,3 | 6,3 | 5031 | 4249 | 4483 |
| 12 | 7,3 | 6,0 | 8,3 | 6941 | 6963 | 7009 |
| 13 | 4,0 | 6,7 | 4,7 | 4392 | 4387 | 4530 |
| 14 | 11,0 | 4,3 | 7,0 | 6399 | 6545 | 6592 |
| 15 | 9,3 | 11,3 | 7,0 | 8722 | 9055 | 8288 |
| 16 | 12,0 | 17,0 | 15,0 | 11108 | 13663 | 13895 |

Table 4.4 shows the difference between the thickness and mass measurements taken before and after the electropolishing process. The Test column represents the order of the tests in the factorial design (Table 3.2). As three replicates were carried out, each test has three results for thickness reduction and mass loss.

## 4.6 - Data processing in *Scilab software*

*Scilab software* was used to carry out the factorial design calculations. A *script* was developed using some of *Scilab*'s statistical tools to set up the mathematical model and graph the influence of each parameter on the response (output) analysed.

The calculations began by assembling the contrast matrix (Table 4.1), which contains all the levels of the interactions and main effects, plus a column representing the overall average of the experiment (it must necessarily be the first column). The contrast matrix is needed to calculate the effects of each parameter and their interactions. Next, the visual (Table 4.3), dimensional (Table 4.4 - Thickness decrease) and mass (Table 4.4 - Mass loss) response matrices were inserted, all including the respective replicates of the tests. The real or original values of the parameters were entered into the *script* for the calculation base.

The analysis of the significance of the main effects and interactions (Table 4.5) was determined using matrix equation 1. Once the numerical value of the effects had been obtained, they were divided by a factor that depends on the planning, in which case the divisor will be $2^4$ for the overall mean and $2^{'41}$ for the other effects (main and interaction).

$$Y_n = X^t y \qquad \text{Eq.(1)}$$

In which:

$Y_n$ is the index of the numerical value of the effect, i.e. the significance value.

X is the contrast coefficient matrix, always with a dimension of k x k/2. y is the results matrix.

XI is the transposed matrix of X.

Table 4.5 - Significance of the effects.

| Parameter to which it refers | Significance for visual analysis | Significance for dimensional analysis | Significance for mass loss analysis |
|---|---|---|---|
| Overall average | 46,0 | 118,3 | 101633,3 |
| C | 10,0 | 24,4 | 202236,7 |
| TC | 15,3 | 35,8 | 35301,3 |
| NC | 10,0 | 33,9 | 32396,0 |
| VT | 0,7 | -0,2 | -2142,0 |
| C.TC | -2,0 | 9,9 | 8040,7 |
| C.NC | -2,0 | 7,5 | 6550,0 |
| C.VT | 2,0 | 0,9 | -900,0 |
| TC.NC | 8,7 | 11,9 | 10741,3 |
| TC.VT | 2,0 | -7,5 | -2223,3 |
| NC.VT | -0,7 | -6,4 | -1786,0 |
| C.TC.NC | -6,0 | 4,6 | 2412,7 |

| C.TC.VT | 0,7 | -6,8 | -1042,7 |
|---|---|---|---|
| C.NC.VT | -2,0 | -1,8 | -782,7 |
| TC.NC.VT | -2,0 | 4,9 | -1304,7 |
| C.TC.NC.VT | -0,7 | 4,9 | -912,0 |

*C - current, CT - cycle time, NC - number of cycles and VT - travel speed.

Table 4.5 shows the significance of each parameter and their interactions with respect to each output analysis. The first column shows the parameters to which each significance refers, and the second row shows the overall average of the tests.

For the visual analysis, i.e. the polishing behaviour at a visual level, Table 4.5 shows that the current, cycle time and the interaction between cycle time and number of cycles account for most of the significance, with 44 of the total. For the dimensional analysis, i.e. the reduction in stent thickness before and after electropolishing, Table 4.5 shows that current, cycle time, number of cycles, the interaction between current and cycle time and the interaction between cycle time and number of cycles have the greatest influence on the dimensional result, totalling 106.8. Finally, for the mass loss analysis, it is notable that current, cycle time, the interaction between current and cycle time and the interaction between cycle time and number of cycles are the most influential effects, adding up to a total significance of 74320.

The standard error of the effects ($s_{efeito}$)must be determined from the square root of the estimate of the joint variance ($s^2$ ) and half the number of trials ($n_n$ ), according to Eq. 2. The estimate of the joint variance will be given by the mean of the variance ($s_1^2$ , $s_2^2$ , $s_3^2$ $s_N^2$ ) for the 2 trials (n ).[k]
trials and by the number of degrees of freedom of the variance ($n_1$ , $n_2$ , $n_3$ ,$n_4$ ,), considering the same number of replicates. The estimate of the joint variance is given by Eq. 3.

$$S_{efeito}=\sqrt{(s^2/n_n)} \qquad \text{Eq. (2)}$$

$$s^2= (v_1 s^2_1+v_2 s^2_2...+v_n s^2_n)/( v_1+ v_2+...+ v_n) \qquad \text{Eq. (3)}$$

Being:

$_{Effect}$ p standard error of the effect.

$s^2$ the variance. Where $s_1$ ,$s_2$ , $s_3$ , ....$s_N$ a for each trial, with its replicates. $n_n$ is half the number of trials, without taking replicates into account, v is the number of degrees of freedom of the variance.

The standard error of the effects of each experiment was 0.353 for the visual analysis, 1.027 for the dimensional analysis (decrease in thickness) and 253.791 for the loss of mass, according to calculations carried out in Scilab following equations Eq.l, Eq.2 and Eq.3.

Finally, the mathematical model for each response was estimated based on the results of the effects calculated for each parameter and parameter interaction. The empirical model that describes the value of the response as a function of the variables under study is given by Response=f(C, TC, NC, VT) and can be estimated by: Rest = bn(l)+ (bn(2)*C) + (bn(3)*TC) + (bn(4)*NC) + (bn(5)*VT) + (bn(6)*(C*TC)) + (bn(7)*(C*NC)) + (bn(8)*(C*VT)) + (bn(9)*(TC*NC)) + (bn(10)*(TC*VT)) + (bn(ll)*(NC*VT)) + (bn(12)*C*(TC*NC)) + (bn(13)*C*(TC*VT)) + (bn(14)*C*(NC*VT)) + (bn(15)*TC*(NC*VT)) + (bn(16)*(C*TC)*(NC*VT)). Where Rest is the estimate of the response and the index bl is the estimator of the overall mean and b2, b3...bl6 are the estimates of the main and interaction effects. These indices can be provided by matrix equation 4.

$$bn=(X^{t}.X)^{-1}X^{t}R \qquad \text{Eq. (4)}$$

Being:

$X^1$ is the transposed matrix of X.

R is the response matrix of the analysed data in question.

The analysis of the planning output in this work operates with three output parameters: visual, dimensional and mass. This resulted in three mathematical models to analyse the influence of the four input parameters. It was decided to use the indices:

- bn - for the model describing the visual analysis;
- dn - for the model describing the dimensional analysis;
- pn - for the model describing the mass loss analysis.

Table 4.6 shows the index values for each output analysis. The values shown here were generated through calculations in *Scilab*.

Table 4.6 - indices for estimating the mathematical model for each output analysed.

| Parameter to which it refers | bn indices (visual analysis) | dn indices (dimensional analysis) | pn indices (mass loss analysis) |
|---|---|---|---|
| Overall average | 2,850 | 7,394 | 6352,083 |
| C | 0,625 | 1,523 | 1264,792 |
| TC | 0,958 | 2,235 | 2206,333 |
| NC | 0,625 | 2,123 | 2024,750 |
| VT | 0,042 | -0,010 | -133,875 |
| C.TC | -0,125 | 0,623 | 502,541 |
| C.NC | -0,125 | 0,469 | 409,375 |
| C.VT | 0,125 | 0,060 | -56,250 |
| TC.NC | 0,542 | 0,748 | 671,333 |
| TC.VT | 0,125 | -0,469 | -138,958 |
| NC.VT | -0,042 | -0,398 | -111,625 |
| C.TC.NC | -0,375 | 0,285 | 150,792 |
| C.TC.VT | 0,042 | -0,423 | -65,167 |
| C.NC.VT | -0,125 | -0,110 | -48,917 |
| TC.NC.VT | -0,125 | 0,310 | -81,541 |
| C.TC.NC.VT | -0,042 | 0,306 | -57,000 |

*C - current, CT - cycle time, NC - number of cycles and VT - travel speed.

Table 4.6 shows the indices for the mathematical models that describe the estimation of the visual, dimensional and mass loss response after the electropolishing process. The first column represents the parameters to which each index is related, with the second line of Table 4.6 representing the overall average of the tests and the others the main effects and their interactions.

Graphs of these mathematical models were also generated for each of the parameters analysed during the electropolishing process. Only the main effects were analysed, i.e. only the input parameters, and not their interactions, since only the main effects generated 12 graphs. Analysing the interactions would have taken too long to interpret, as it would have generated a larger number of graphs. Figures 4.6, 4.7 and 4.8 below show the graphs generated from the mathematical models created to analyse the responses to the factorial design. The four graphs of the main effects at their maximum (1) and minimum (-1) levels are shown on the abscissa axis and the estimate generated by the mathematical model on the ordinate axis.

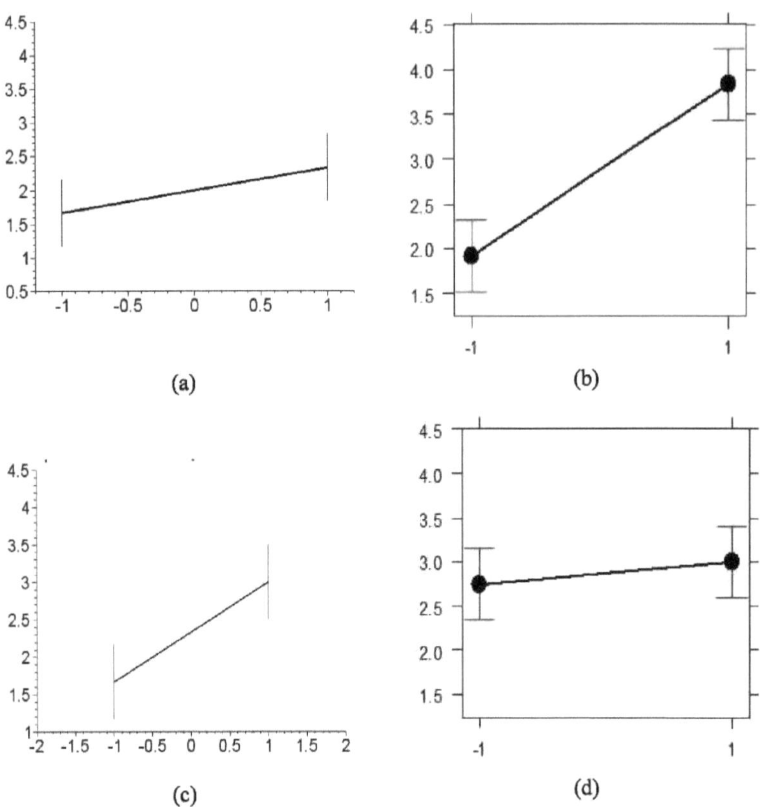


Figura 4.6 - Influence of current (a), cycle time (b), number of cycles (c) and speed of speed (d) on the visual analysis.

Figure 4.6 shows the influence of current (a), cycle time (b), number of cycles (c) and translation speed (d) for analysing the external surface (visual analysis) of the electro-polished *stents*. These graphs were generated from the input data in the *script* and the linear regression analysis for modelling the mathematical model. It is notable that the parameters current, cycle time and number of cycles have the greatest influence on the visual analysis, since graphs (a), (b) and (c) in Figure 4.6 have a steeper slope than graph (d), which represents the translation speed.

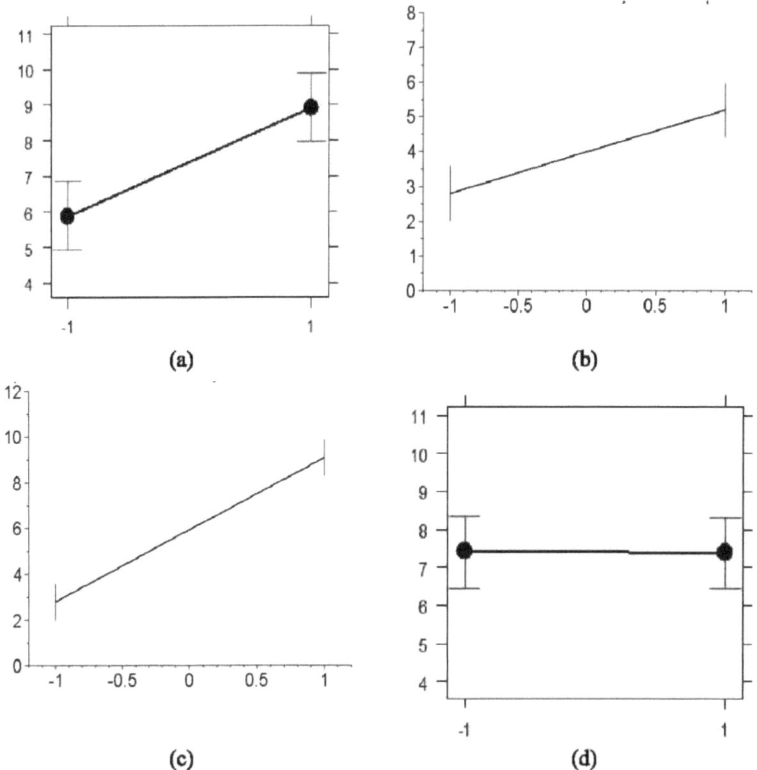


Figura 4.7 - **Influence of current (a), cycle time (b), number of cycles (c) and translation speed (d) on dimensional analysis.**

**Figure 4.7 shows the influence of current (a), cycle time (b), number of cycles (c) and translation speed (d) when analysing the reduction in thickness of electropolished *stents* compared to *stents* before undergoing this process. Analysing Figure 4.7 once again shows the slope of the first three graphs (a, b and c - current, cycle time and number of cycles, respectively), demonstrating the influence of these parameters on the dimensional analysis as well. The graph for translation speed (d) shows practically no slope, showing that this parameter has little influence on the final thickness of the electro-polished *stent*.**

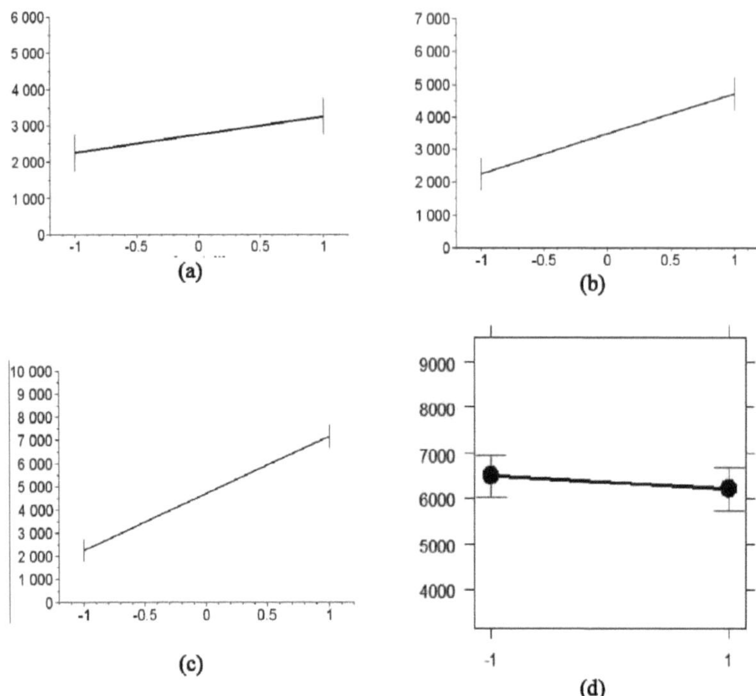


Figure 4.8 - Influence of current (a), cycle time (b), number of cycles (c) and travelling speed (d) on mass loss analysis.

Figure 4.8 shows the influence of current (a), cycle time (b), number of cycles (c) and translation speed (d) when analysing the mass loss of electropolished *stents* compared to *stents* before undergoing this process. Once again, the influence of current, cycle time and number of cycles on the contribution of *stent* mass loss *can be* seen. The translation speed (Figure 4.8d) once again shows a slight slope, this time decreasing. This shows that increasing the speed helps to reduce mass loss.

# CHAPTER 5

## CONCLUSIONS

By analysing the 2-factor design[4] carried out, the influence of the chosen parameters (current, cycle time, number of cycles and translation speed) on the change in gloss of the outer surface of electropolished *stents*, the reduction in their thickness and the loss of mass generated by the process can be understood. Both the significance of each parameter and their interactions were observed and the response estimated using the *script* and graphs generated in *Scilab.*

For the visual analysis, i.e. the polishing behaviour at a visual level (considered poor - 1, intermediate - 3 and adequate - 5) it can be seen from Table 4.5 and Figure 3.6 that the current, cycle time and number of cycles have the greatest influence on the external surface integrity of electropolished *stents*. An increase in these parameters leads to an increase in the level of polishing.

For the dimensional analysis, i.e. the reduction in *stent* thickness before and after electropolishing, it can be seen that once again the current, cycle time and number of cycles have a greater influence on the analysis. It can be seen that the translation speed had little influence on the dimensional analysis, as well as the visual analysis.

Finally, when analysing mass loss, it is possible to see that current and cycle time have the greatest influence. An increase in these parameters leads to an increase in the mass loss of electropolished *stents*. The translation speed showed a downward slope (Figure 4.8d), demonstrating that increasing it generates a tendency to reduce mass loss during the process.

According to Aihara (2009), increasing the current and time of electropolishing reduces the roughness of the electropolished surface. This is because the current is responsible for removing ions from the metal to be polished, which are deposited on the negative pole of the process, and the longer the time during which electropolishing takes place, the less rough the layer will be. The tests carried out show the influence of these two parameters during the electropolishing process. Even though the roughness itself was not measured, the reduction in thickness and the better level of polishing served as the basis for this proof.

The number of cycles in this case influences the total polishing time. According to the way the electro-polishing machine works, *it is* possible to carry out a cycle, for example of 10 seconds, and repeat it as many times as necessary. As the number of cycles is linked to time, once again we see proof of its influence on electro-polishing. One point to emphasise is that the parameter of number of cycles together with cycle time, i.e. their interaction, should be better investigated since the same total time can be obtained with values of number of cycles and cycle time in different combinations.

The translation speed of the electrode tower (Figure 3.3) had very little influence on the response analyses. This parameter itself does not seem to have any significant direct influence on the responses analysed, nor when we look at its interactions with the other parameters. However, the practice of the electropolishing process showed subtle differences in the external surface integrity of the *stents* when the speed of translation was changed.

It can be concluded that the factorial planning tool proved to be very effective for processing the data and studying the influence of the parameters on the electropolishing process studied in this work. The randomisation of the tests, their statistical analysis and the possibility of generating an empirical model that describes the behaviour of an output variable to be analysed on the basis of controllable input variables, provide relatively short-term development when it comes to research, whether in an academic or industrial environment. By setting up and carrying out a factorial design, it was possible to understand the behaviour of the external surface of coronary *stents* after they had been electropolished and the influence of current, cycle time, number of cycles and translation speed (controlled input variables) on the responses analysed. The surface quality, thickness and final weight of an electropolished *stent* depend directly on the current, cycle time and number of cycles. The influence of interactions between these variables can be studied in more depth in future work.

The *Scilab software* also proved to be an appropriate tool for carrying out the calculations intrinsic to factorial planning, since it would be very labour-intensive to carry out calculations on matrices with dimensions of 16x16, for example, as was the case with the calculations carried out during the course of this work.

# CHAPTER 6

## BIBLIOGRAPHICAL REFERENCE


ABELIN, A. P.; QUADROS, A. S.; ZANETTINI, M. T.; LEBOUTE, F. C.; YORDI, L. M.; CARDOSO, C. R.; MORAES C. A. R; RODRIGUES, L. H. C.; JÚNIOR, L. H. C. R.; MOURA, M. R. S.; SARMENTO-LEITE, R.; GOTTSCHALL, C. A. M. Twelve years of experience with coronary stent implantation in 5284 patients. **Revista Brasileira de Cardiologia Invasiva,** 17, 3,346 -51,2009.

AIHARA, H. **Surface and biocompatibility study of electropolished Co-Cr alloy L605.** 2009. Dissertation (Master's in Engineering) - Department of Engineering, São José State University, São José.

**ANDRAMED U-FLEX PERIPHERAL SELFEXPANDABLE NITINOL STENT.** MedintPro. Available at: <http://www.medint.sk/andramed-u-flex-peripheral-selfexpandable-nitinol-stent/>. Accessed on: 02 October 2014.

ARAÚJO, R. **Numerical simulation of the expansion process of stents for angioplasty by hydroforming.** 2007. Dissertation (Master's in Engineering) - Department of Engineering. Federal University of Uberlândia. Minas Gerais.

CHAMIÉ, D.; ABIZAID, A. Is it time to adopt a national stent? **Brazilian Journal of Invasive Cardiology,** 17, 3, 300-4, 2009.

COSTELLA, R. S. **Evaluation and surface treatment of Co-Cr ARTM F90 superalloy stents.** 2010. Diploma work - Department of Engineering. Federal University of Rio Grande do Sul. Rio Grande do Sul.

CUNICO, M. W. M., MIGUEL, O. G., ZAWADZKI, S., F., PERALTA-ZAMORA, P., VOLPATO, N., Factorial Planning: a valuable statistical tool for defining experimental parameters used in scientific research. **Visão Acadêmica,** Curitiba, v.9, n 1, p 23 - 32, jan/jun 2008.

FILHO, R. G. A., **Fractional factorial designs for sensitivity analysis of simulation models and events.** 2006. Dissertation (Master's in Production Engineering) - Department of Engineering. Federal University of Itajubá. Minas Gerais.

GALDAMEZ, E. V. C., **Application of the techniques of planning and analysing experiments to improve the quality of a manufacturing process for plastic products.** 2002. Dissertation (Master's in Engineering) - Department of Engineering, University of São Paulo, São Paulo.

MARINHO, M. R. M.; CASTRO, W. B., Factorial planning: a powerful tool for researchers. In: COBENGE, XXXIII, 2005, Paraíba. **Electronic annals.** Available at: <http://www.abenge.org.br/cobenges-anteriores/2005/2005-xxxiii-cobenge-campina-grande-pb>. Accessed on: 01 August 2014.

MOREIRA, P. N. T., **Planning and optimisation of a chemiluminescent method for the determination of vitamin B12 using a flow-batch system.** 2009. Dissertation (Doctorate in Chemistry) - Department of Chemistry. Federal University of Paraíba.

Jõao Pessoa, Paríba.

OLIVEIRA, A. C. S., **Microstructural and hardness analysis of heat-treated ASTM F90 Haynes® 25 (L605) Co-Cr alloy tubing for medical devices.** Diploma work (Graduation in Materials Engineering) - Department of Engineering, Federal University of Rio Grande do Sul, Porto Alegre, 2010.

SCHAULSOHN, N., **ASTM F2516 Tension Testing of Nitinol How to Guide.** Available at: <http://info.admet.com/specifications/bid/54288/ASTM-F2516-Tension-Testing-of-Nitinol-How-to- Guide> . Accessed on: 02 October 2014.

**Vascular devices -** *stent* **machining.** Norman Noble. Available at <http://www.nnoble.com/MedicalDevices/Vascular/Vascular.htm>. Accessed on: 01 August 2014.

VEROLI, A. B., **Study of galvanostatic electropolishing of AISI 304 stainless steel using concentrated acid solutions.** 2011. Dissertation (Master's in Chemistry) - Department of Chemistry, Federal University of São Carlos, São Paulo.

yes

# I want morebooks!

Buy your books fast and straightforward online - at one of world's fastest growing online book stores! Environmentally sound due to Print-on-Demand technologies.

Buy your books online at
**www.morebooks.shop**

---

Kaufen Sie Ihre Bücher schnell und unkompliziert online – auf einer der am schnellsten wachsenden Buchhandelsplattformen weltweit! Dank Print-On-Demand umwelt- und ressourcenschonend produzi ert.

Bücher schneller online kaufen
**www.morebooks.shop**

info@omniscriptum.com
www.omniscriptum.com

Printed by Books on Demand GmbH, Norderstedt / Germany